To my parents, who instilled in me a love for learning. To my husband, James, and our children, Jillian, Jayla, and James Shawn—you inspired every word of this book. I wrote this for you. All my love. —S.D.F.

For Audrey Elizabeth, the smartest, most hard-working, and sweetest person I know —M.L.A.

Room to Read would like to thank Tatcha™ for their generous support of the STEAM-Powered Careers collection.

Copyright © 2022 Room to Read. All rights reserved.

Written by Stacey Finley
Featured scientist: Stacey Finley
Illustrated by Michelle Laurentia Agatha
Edited by Carol Burrell
Photo Researcher: Kris Durán
Series Art Director and Designer: Christy Hale
Managing Editor: Jamie Leigh Real
Series Editors: Carol Burrell, Jamie Leigh Real, Jocelyn Argueta, and Deborah Davis
Series Creators: Jocelyn Argueta and Dr. Dieuwertje "DJ" Kast
Copyeditors: Debra Deford-Minerva and Danielle Sunshine

ISBN 978-1-63845-062-7

Manufactured in Canada.

10 9 8 7 6 5 4 3 2

Room to Read
465 California Street #1000
San Francisco, California 94104
roomtoread.org

World change starts with Educated children.©

DATA SCIENCE

by **Stacey Finley**, featured scientist

illustrated by **Michelle Laurentia Agatha**

contents

Explore Data Science with Cora and Bonnie   6

What Is Data Science?   22

Meet the Scientist   24

Learn More about Data Science   30

Word List   34

Cora grabs the clubhouse laptop and slides it into her backpack. She has a look of excitement and wonder on her face.

"What are you thinking about?" asks her friend Bonnie, fluttering around Cora's head.

"I just heard of something so amazing that I have to go and

check it out for myself! Want to come?" Cora replies.

"Yay! I love adventures!" says Bonnie. "Where are we going?"

"Well," says Cora, "I heard that data science can help people make better decisions about, well, everything—from medicine to the environment. So we're going to a special research laboratory that focuses on data science!"

"This should be interesting!" says Bonnie. "Let's go!"

Data Science

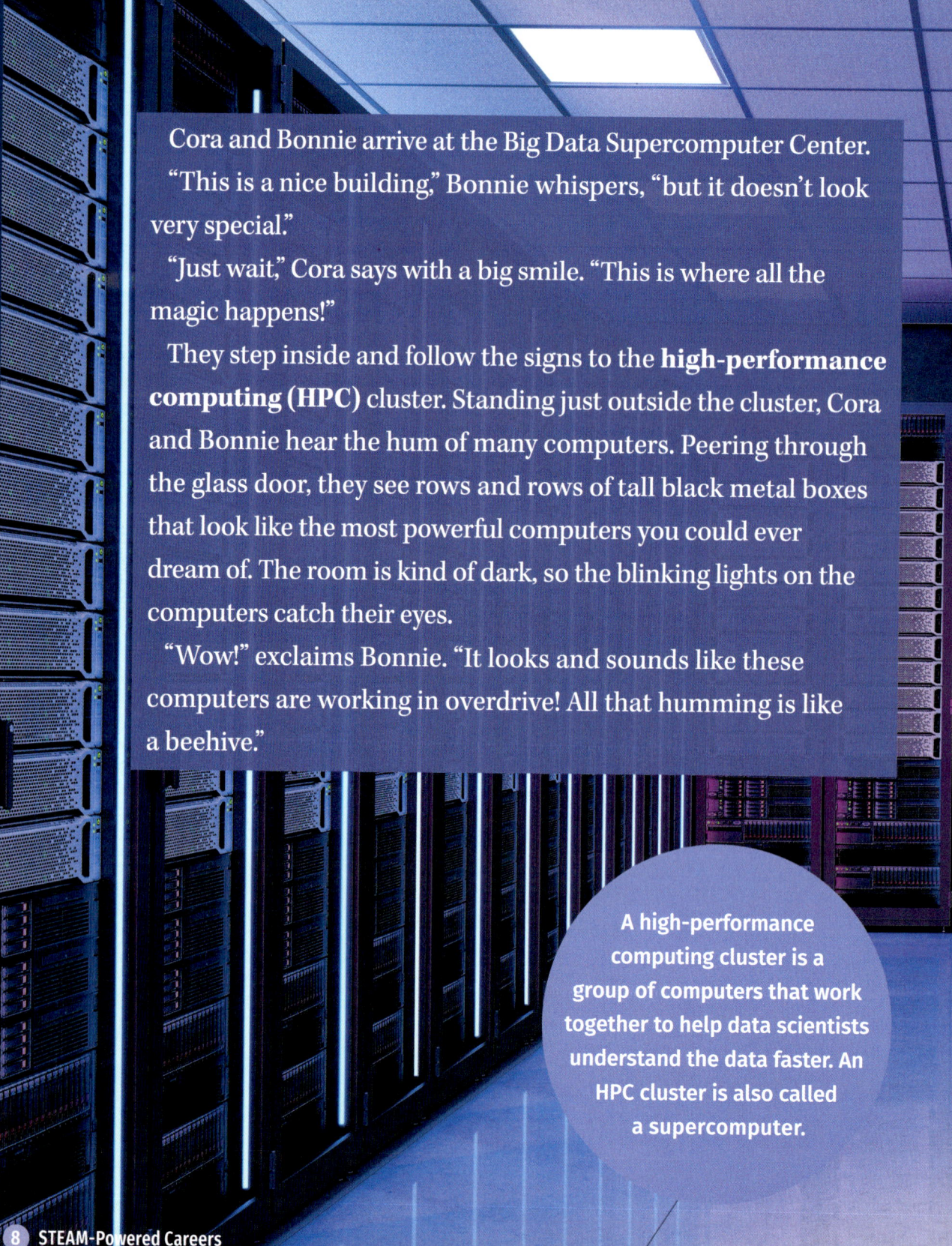

Cora and Bonnie arrive at the Big Data Supercomputer Center.

"This is a nice building," Bonnie whispers, "but it doesn't look very special."

"Just wait," Cora says with a big smile. "This is where all the magic happens!"

They step inside and follow the signs to the **high-performance computing (HPC)** cluster. Standing just outside the cluster, Cora and Bonnie hear the hum of many computers. Peering through the glass door, they see rows and rows of tall black metal boxes that look like the most powerful computers you could ever dream of. The room is kind of dark, so the blinking lights on the computers catch their eyes.

"Wow!" exclaims Bonnie. "It looks and sounds like these computers are working in overdrive! All that humming is like a beehive."

A high-performance computing cluster is a group of computers that work together to help data scientists understand the data faster. An HPC cluster is also called a supercomputer.

STEAM-Powered Careers

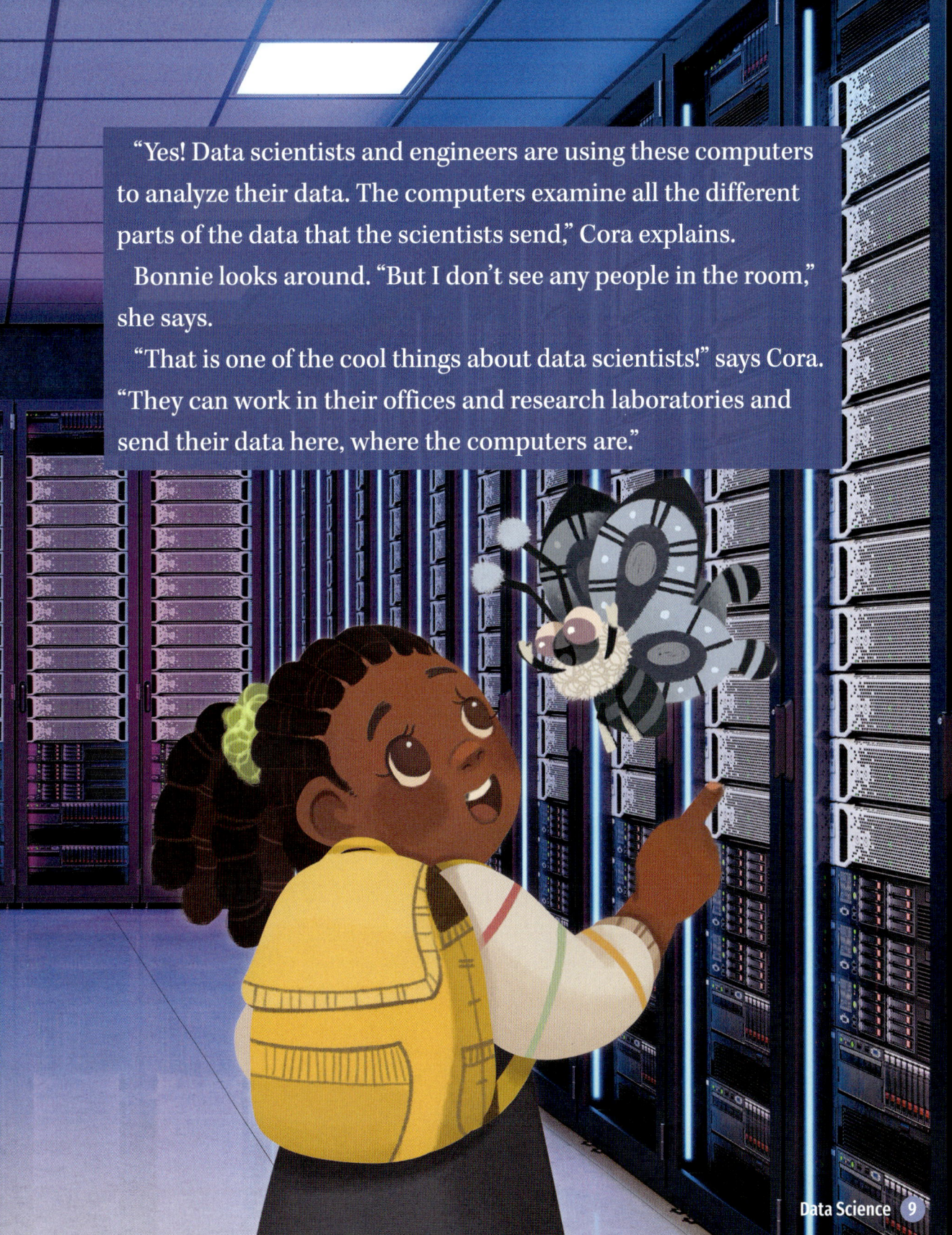

"Yes! Data scientists and engineers are using these computers to analyze their data. The computers examine all the different parts of the data that the scientists send," Cora explains.

Bonnie looks around. "But I don't see any people in the room," she says.

"That is one of the cool things about data scientists!" says Cora. "They can work in their offices and research laboratories and send their data here, where the computers are."

"I have so many questions!" Bonnie says.

> What is data? How does it get here? What do the data scientists use the data for?

"Slow down, Bonnie!" Cora replies. "These are great questions. And I have the perfect way to find out more! You're going to love this!"

Cora opens the door to the HPC cluster and pauses dramatically. "Follow me!" she calls. She swings her arms around a few times and then, with a huge leap, she jumps straight into the computer and disappears!

To Bonnie's surprise, thousands of little zeros and ones splash out of the computer like tiny raindrops! A few little sparkly blue numbers fall on her wings.

Bonnie flies up to the computer.

With a *GULP*

    and a deep breath,

        Bonnie dives right in.

"Wow! We are inside a computer!" exclaims Bonnie.

"Yes! Isn't it cool?" Cora says. "We are right in the middle of where a computer stores data and does calculations."

"But what are all these numbers?" asks Bonnie.

Cora explains, "The purpose of a computer is to process data, such as a set of numbers. The computer turns data into information—something we can understand."

Bonnie looks confused.

Cora continues, "OK, you know how I love to make fruit smoothies? We put the strawberries, bananas, yogurt, and ice into the blender, which is also called a food processor. The food processor does its job, and out comes a smoothie!" says Cora.

"Oh, yeah! They're so yummy!" Bonnie replies.

"Well," says Cora, "it helps to think of a computer like a *data* processor. It takes in data, holds it, and chews on it for a bit. Then the computer pours out the results."

There are names for each "job" a computer does. The information that goes in is called **input**. Computers store information in their **memory**, just like your brain does! When computers do things with the information, it is called **processing**. The computer's results are the **output**.

Data Science

"Wow, now I get it!" says Bonnie. "But where do the strawberries and bananas—I mean data—come from?"

"A data scientist sends their data to the computer clusters, right from their office!" Cora replies.

"What are data scientists?" Bonnie asks.

"Data scientists are people who are experts at looking at and understanding data. They can use data to solve all kinds of problems," Cora explains.

"Like what?" asks Bonnie.

"Well, I heard about a data scientist who uses data to help doctors figure out how to help their patients feel better."

Bonnie is amazed. "That sounds important!"

"Yes," says Cora. "Data scientists also help companies make better products. Products that use something called **machine learning** can even help people like my parents find the fastest way to get to our favorite ice cream place!"

Bonnie's mouth drops open in surprise.

"Let's get out of this computer and go see where a data scientist works," Cora says excitedly.

Machine learning uses data to train computers to do things without needing a human to tell them the exact steps to follow. One example is the self-driving car! Machine learning helps people make decisions and helps companies work better.

Data Science

Cora shows Bonnie a nice office with a comfortable chair. "This is where a data scientist can do their hard work!"

Bonnie flies around to look at everything on the desk. "More computers?" she asks.

"Yes!" Cora says, beaming. "The data scientist uses these computers to make a **mathematical model**."

"What's a mathematical model?" Bonnie asks.

"A model is something that tries to imitate a real thing," says Cora. "And a mathematical model uses math to imitate the real thing."

One area of data science is **mathematical oncology**. Data scientists in this area use data and mathematical models to help doctors who study cancer. The data scientists can use the models to predict the best way to help patients.

Cora grins at Bonnie. "Guess what?"

"What?" Bonnie says.

"*I* made a model!" Cora says proudly. "And you were there to see it!"

Bonnie waggles her antennas. "Really? When?"

"Remember how I got the building blocks for my birthday to build a pretend ice cream shop? That little ice cream shop was a model of a real ice cream shop."

"Yummy . . ." says Bonnie, thinking about ice cream again.

"A mathematical model uses data so it can be as close to the real thing as possible. Then the data scientist uses the model to make real-life predictions! Making a prediction means to give an idea of what will happen in the future."

"And how does the data scientist use the HPC cluster you just told me about?" asks Bonnie.

"Mathematical models can get very complicated. So the data scientist uses the powerful computers at the HPC cluster to get the right answers," Cora explains.

"OK, Bonnie, there's one more important thing I have to tell you about data scientists," Cora announces with a smile. "To make mathematical models and help solve problems, they use **computer languages**."

"Wait, they speak different languages?" Bonnie asks, confused again.

There are different computer languages that we can use to tell a computer what to do. Just like there are many languages that people use to talk with one another. Some common computer languages are Python, Java, C++, and PHP.

"Not exactly," says Cora. "But a data scientist has to tell a computer what to do. The computer only works if the data scientist uses a language it can understand."

"Can I learn a computer language?" asks Bonnie.

"Yes!" Cora says. "Anyone can learn. Learning computer languages helps you become a data scientist!"

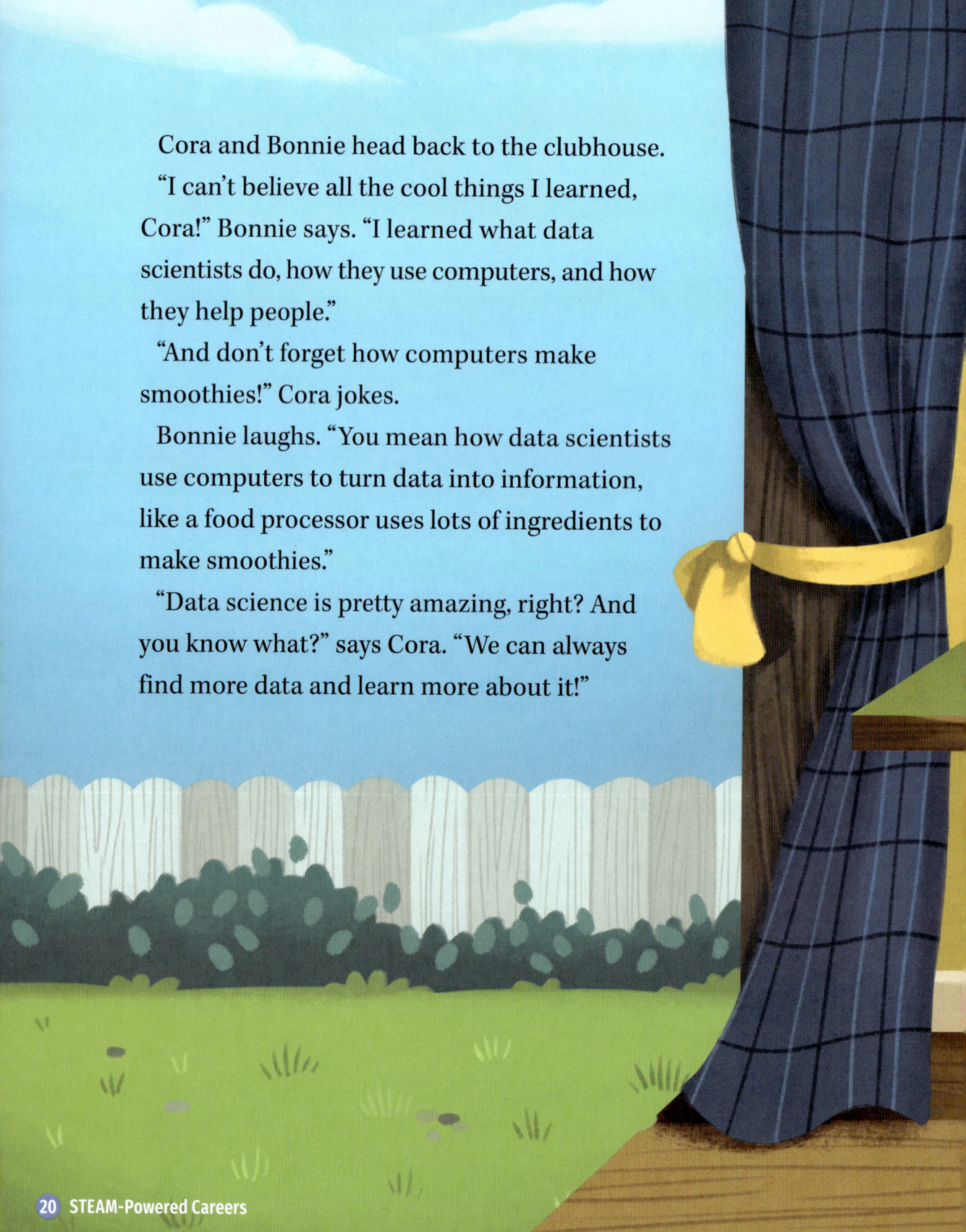

Cora and Bonnie head back to the clubhouse. "I can't believe all the cool things I learned, Cora!" Bonnie says. "I learned what data scientists do, how they use computers, and how they help people."

"And don't forget how computers make smoothies!" Cora jokes.

Bonnie laughs. "You mean how data scientists use computers to turn data into information, like a food processor uses lots of ingredients to make smoothies."

"Data science is pretty amazing, right? And you know what?" says Cora. "We can always find more data and learn more about it!"

Data Science 21

What is Data Science?

Cora and Bonnie just gave us a lot of data to process! To help analyze all of that, **Dr. Stacey Finley** will give us an inside look at what it's like to be a data scientist. First, let's go over some of the information.

Data science is an area of study that deals with a lot of—you guessed it—data!

By using computers and math, data science helps us find patterns in the data. It turns data into something we can understand. We can use this information to make better, more informed decisions.

Dr. Stacey Finley works with students all around the world and uses data to help find new ways to treat diseases like cancer. Let's ask her some questions and make sure we've got some new and interesting information coded in our brains.

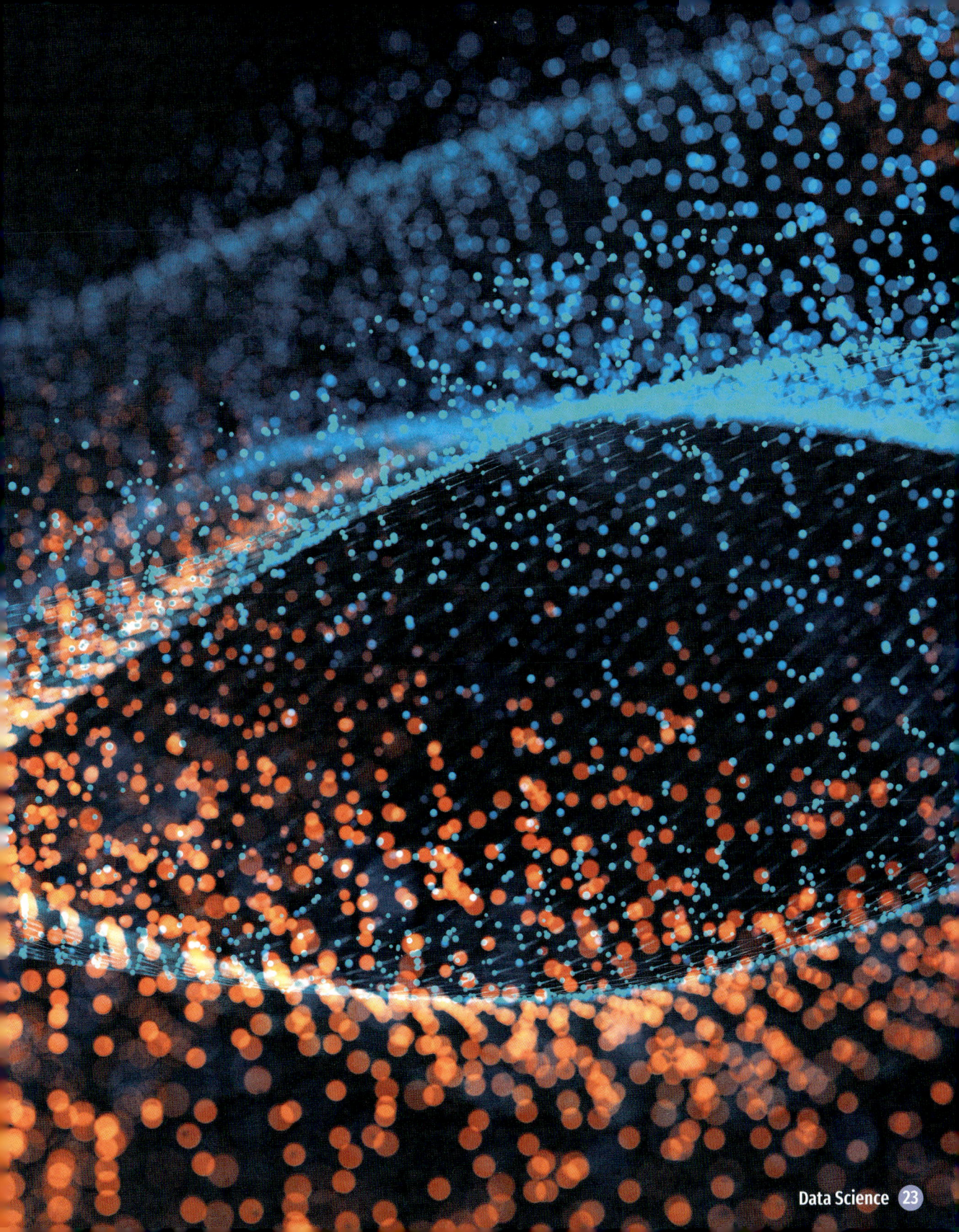

Meet the Scientist
Dr. Stacey Finley

I studied chemical engineering in college to receive my undergraduate degree, and then I earned a PhD, also in chemical engineering. For my undergraduate degree, I went to an HBCU, a historically Black college or university—a school created when Black students were not allowed to go to other schools. Being around Black students and professors gave me a lot of confidence to become a professor myself.

Fun Fact #1: When I am not working, I spend a lot of time with my family—my husband and three children. We love to be outdoors.

Fun Fact #2: I have run seven marathons, all around the country, including in Chicago, Indianapolis, and Washington, D.C.

What is your favorite thing about data science?

What is your least favorite thing about data science?

It is really cool that I get to answer questions about the world using computers. Computers help us work really fast and get answers that help people.

Some people think that data science is too hard to learn. I would love for everyone to know that data science is hard, but it is also fun and rewarding, and it helps us understand things around us.

Come with me and I'll show you my daily routine!

Data Science 25

A Day in the Life, Part 1

I start almost every day with a workout because I like to stay healthy. Sometimes I lift weights or run outside.

I am a professor at the University of Southern California (USC). My office is in a building where there are lots of other scientists who work on different projects. We get to talk about fun science and learn from each other.

I am in my office most days, working on my computer. I use the computer to make new mathematical models of cancer. I usually listen to music while I work, including classical piano music.

I spend a lot of time talking to my students. I walk over to their work area every day just to hear about what they are working on.

I help my students think of solutions for their projects and how to use data science in their research. This is one of the things I love most about my job!

It is exciting to see my students work on a problem for several years and learn so much. I am proud when they graduate!

Data Science

A Day in the Life, Part 2

Another important part of being a professor is teaching. I teach a class about building mathematical models using data.

Sometimes I host visitors on campus and tell them about our research in mathematical modeling.

I talk to kids about data science, modeling, and engineering. I encourage them to ask lots of questions and learn about science.

$2a = b + 3$

?

I also get to travel to colleges all around the world and talk about my work. This is a chance to let other researchers know what I have learned and to hear about new ideas.

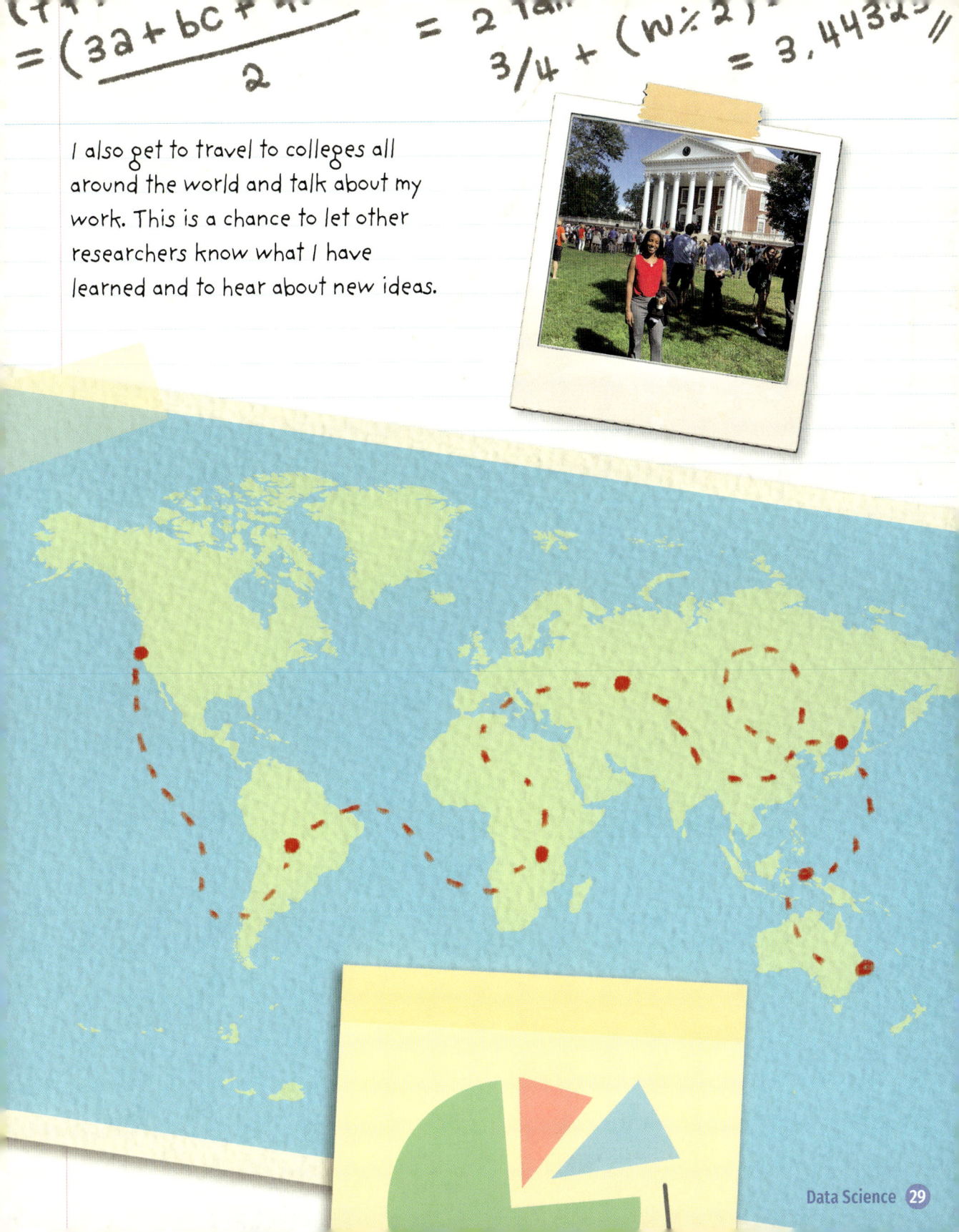

Data science is an important part of careers in many fields!

STEAM Careers in Data Science

People in many fields, from computer programming to medicine, use data science every day. It's also used in statistics, the branch of mathematics that works with collecting, analyzing, and interpreting data, and engineering, as problem-solvers come up with and build solutions to challenging problems.

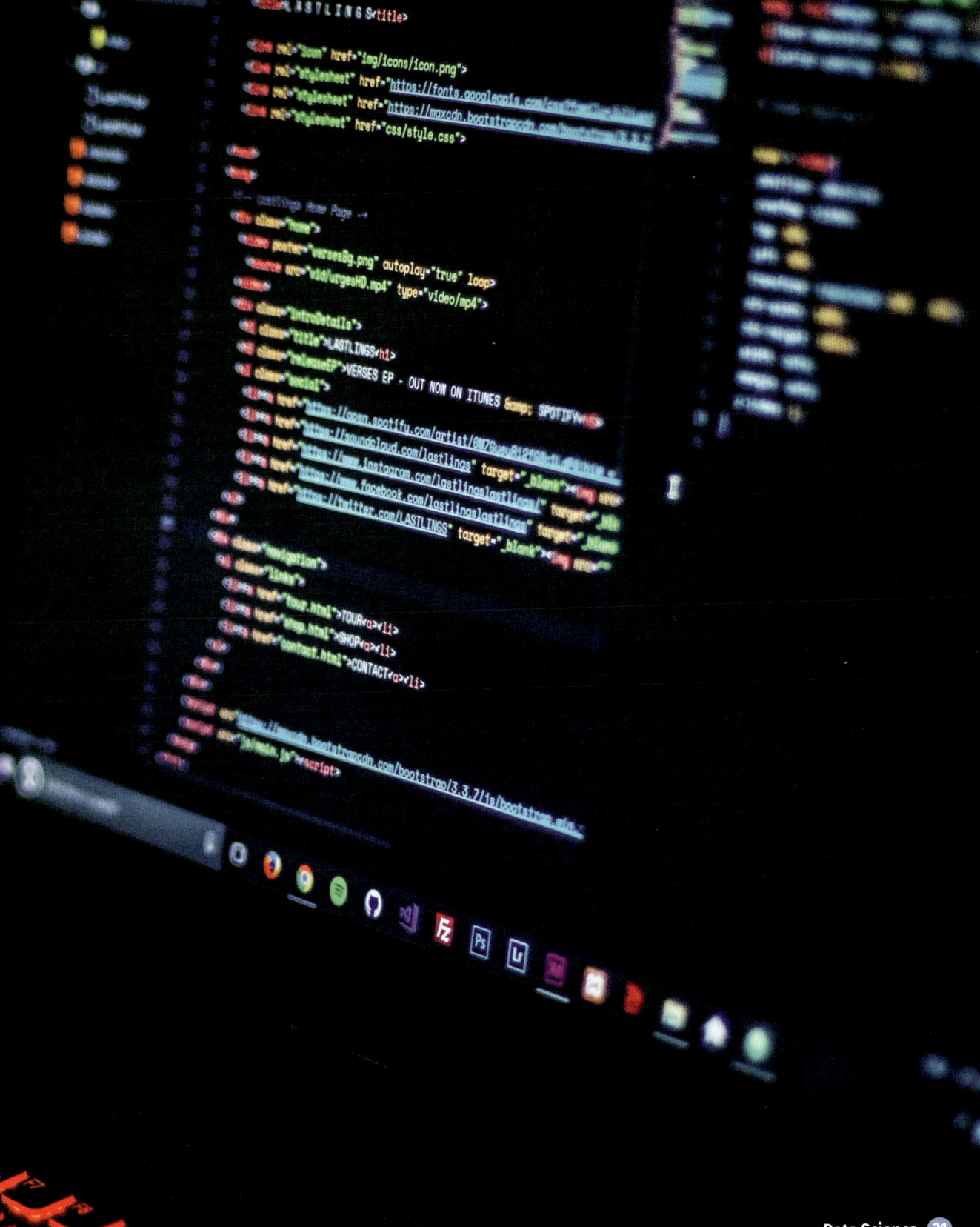

Data Science

The Future of Data Science

In the near future, data scientists will be focusing on two areas: tools and applications. They will ask a lot of questions and look for a lot of answers!

Tools

Can we develop new tools and approaches to make data analysis and machine learning faster and more efficient?

How can we train computers to do machine learning when only a little data is available? How can we help data scientists around the world share their data to make better predictions?

Applications

How can we apply data science to improve the world around us?

How can doctors decide on the best treatment for a patient? Can we use data science and wearable devices (like a smartwatch) to predict when someone will get sick?

How do we design safe self-driving cars?

How can we use data science to provide energy at the right time to the right places?

Can we use machine learning and artificial intelligence to help robots communicate with people who are sick or have disabilities?

Do You Want to Be a Data Scientist?

You can start now by doing all of these things:

- **Ask lots of questions.** Try to figure out how things around you work.
- **Learn how to code.** This will let you do fun things like create a video game or make a robot follow your commands.
- **Use math and numbers in your everyday life.** Brainstorm ways that you can use math in regular activities. For example, math is part of baking! We have to think about numbers and fractions when following a recipe. Accidentally swapping two cups of flour for two teaspoons of salt would ruin supper!

Word List

binary numeric system: how data looks inside computers, using only zeros and ones

computer language: a language people use to write instructions for a computer

high-performance computing (HPC) cluster: a group of computers that work together to help data scientists understand data faster. Also called a supercomputer.

input: information put into a computer

machine learning: when a computer uses data to train itself to do things without a human to tell it the steps to follow

mathematical model: when data is interpreted using math so a model or prediction can be made that is as close to the real thing as possible. A weather forecast is an example of a mathematical model.

mathematical oncology: an area of data science that uses *mathematical models* to help doctors who study cancer

memory: where computers (and humans and butterflies) store information

output: the processed information that a computer shares

processing: when a computer uses code to go through *input*, and then translates it into something we can read as *output*

Data Science Resources

Check out these books:

The ABCs of Data Science, created for future data scientists by current data scientists
   https://theabcsofdatascience.com

You can read this as an e-book or a real-life book:

Florence the Data Scientist and Her Magical Bookmobile by Ryan Kelly, illustrated by Mernie Gallagher-Cole
   https://www.dominodatalab.com/childrens-book-florence-the-data-scientist

Acknowledgments

University of Southern California

Viterbi School of Engineering; Dornsife College of Letters, Arts, and Sciences

Biomedical Engineering; Quantitative and Computational Biology; Mork Family Department of Chemical Engineering & Materials Science

Dr. Stacey Finley, PhD, is an associate professor at the University of Southern California in Los Angeles. She directs a research group to develop new mathematical models that help us understand how tumors grow. Dr. Finley also teaches in the USC Viterbi School of Engineering. She is passionate about sharing her excitement about math and how math is part of the world around us.

Michelle Laurentia Agatha was born in Jakarta, Indonesia. Ever since she was young, she has had a huge interest in cartoons and illustrated books. Michelle pursued her dream of becoming an illustrator by earning a Bachelor of Fine Arts degree from the Academy of Art University in San Francisco. Currently, Michelle is working as a children's book illustrator, concept artist, and UI/UX designer.

Explore the Complete

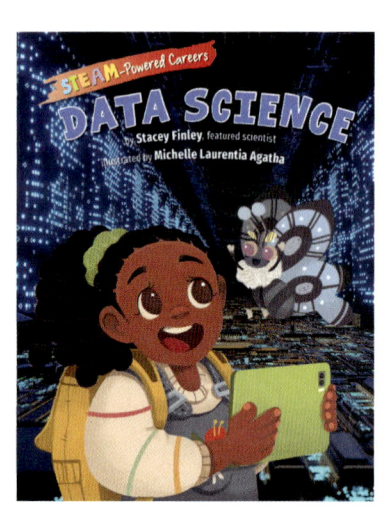
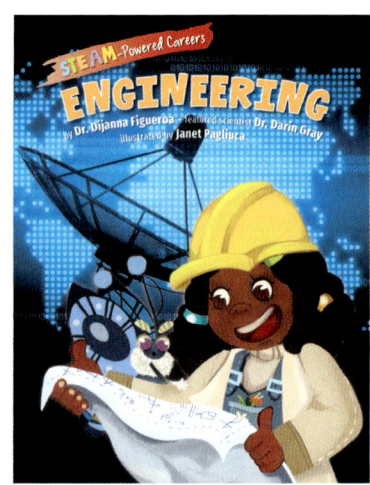

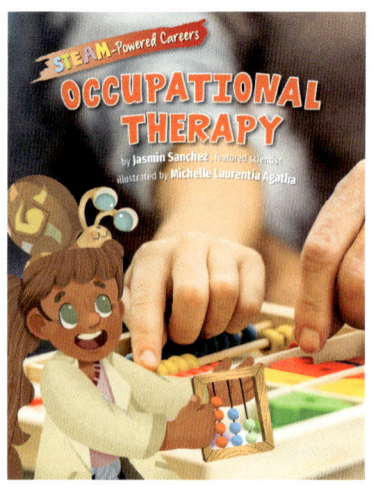
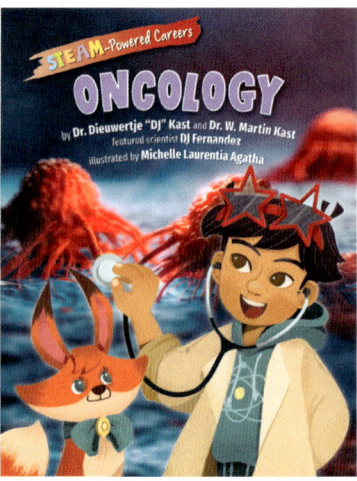

STEAM-Powered Careers Series!

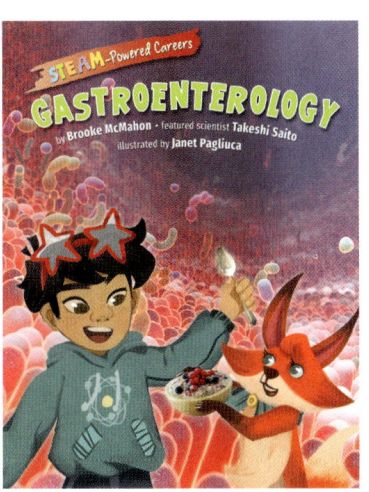

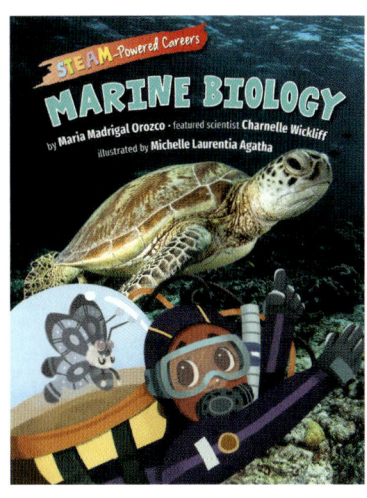
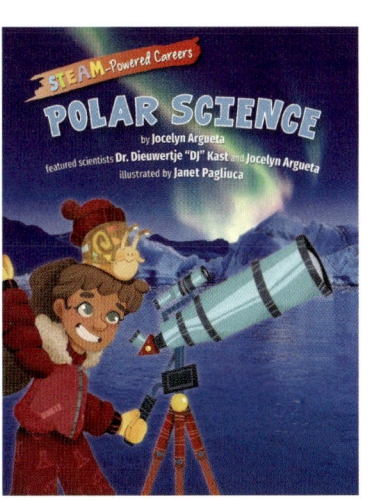
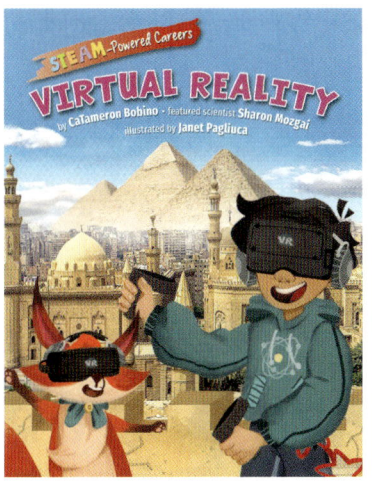

Photo Credits

Cover iStock.com/amynapaloha **6–7** photo by Markus Spiske from Pexels; photo by Christina Morillo from Pexels; iStock.com/monsitj **8–11** iStock.com/onurdongel **12–13** iStock.com/luismmolina; iStock.com/PhonlamaiPhoto; Pexels/Pixabay **14–15** iStock.com/Blue Planet Studio; iStock.com/juststock; iStock.com/Prostock-Studio; iStock.com/sweetym **16–17** photo by Tranmautritam from Pexels; © Didecs | Dreamstime.com; iStock.com/unpict **19** iStock.com/kali9 **22–23** iStock.com/piranka **24–25** Tiffany J Photography; photo courtesy of Stacey Finley; Jess M. Christopher; Gus Ruelas **26–27** photos courtesy of Stacey Finley; **28–29** Gus Ruelas, USC Dornsife/Marc Merhej, photos courtesy of Stacey Finley; Photo by Pixabay from Pexels **30–31** Gus Ruelas; photo by Danny Meneses from Pexels **32–33** kanawatvector/Depositphotos.com **34–35** iStock.com/in-future **36–37** kyolshin/Depositphotos.com; Tiffany J Photography; photo courtesy of Michelle Agatha **40** Victorgrigas, CC BY-SA 3.0, via Wikimedia Commons